AF230886

4th Grade Math
Volume 2

© 2013 OnBoard Academics, Inc
Newburyport, MA 01950
800-596-3175
www.onboardacademics.com

ISBN: 978-1-939796-83-7

Table of Contents

Order of Operations

Key Vocabulary

order of operations

parentheses

exponents

3 + 4 x 2 = ?

The first operation is in red. Look at what a difference it makes when we change the order of operations.

3 + 4 x 2
= 7 x 2
= 14

3 + 4 x 2
= 3 + 8
= 11

Parentheses

By adding parentheses ( ) you can help to make sure the order is clear.

Look at the order of operation below. Always work from left to right, calculate what is in parentheses first. Multiplication is first, then the division, then the addition and then the subtraction always working left to right.

(3 + 4) x 2 = 14

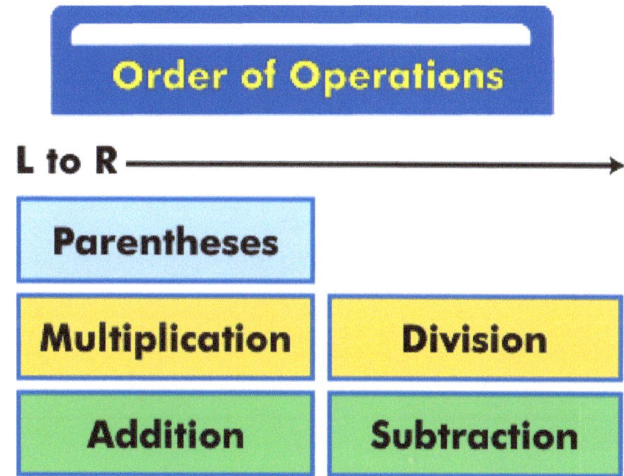

An Example of the Order of Operations
The operation for each step is in red.

Try this problem using the example above to guide you.

1	$42 \div 3 - (4 + 2)$	$42 \div 3 - 6$	**Parentheses**
2	$42 \div 3 - 6$	$14 - 6$	**Division**
3	$14 - 6 = 8$		**Subtraction**

$$52 \div 4 - (15 - 8)$$

Practice

1	**4 x (3 + 5) =**
2	**(6 − 2) x (7 + 2) =**
3	**10 − 5 x 2 =**
4	**35 ÷ 5 − 4 =**

Draw in the parentheses to make these statements true.
The first one is completed for you

1	$(8 \div 4) + 2 = 4$
2	$7 - 2 \times 5 \times 3 = 75$
3	$5 + 4 \times 6 + 2 = 72$
4	$8 - 3 \times 9 - 1 = 44$

Name_________________________________

Order of Operations Quiz

1   **True or false? 7 + 8 x 3 - 2 = 43**

2   **Which of these equations is *not* correct?**

    A   **6 + 7 x 8 = 62**

    B   **8 + 7 x 6 = 90**

    C   **6 x (7 + 8) = 90**

    D   **6 x 7 x 8 = 336**

3   **Use order of operations to solve 9 ÷ 3 x 4 + 12 = ?**

4   **Use order of operations to solve 9 ÷ 3 x (4 + 12) = ?**

Number Properties

Key Vocabulary

distributive property

commutative property

associative property

zero property

Does order matter? ____

____

Alison sprinkled pepper and cheese on her pizza.

Owen sprinkled cheese and pepper on his pizza.

Fernando had a shower and then put on his PJs.

James put on his PJs and then had a shower.

True or false?

1. $3 + 4 = 4 + 3$

2. $5 \times 6 = 6 \times 5$

3. $8 - 5 = 5 - 8$

4. $10 \div 5 = 5 \div 10$

The commutative property of addition and multiplication

> **The commutative property of addition and multiplication:
> when you add or multiply numbers, you can swap the
> order of the numbers and still get the same answer.**

Discover the commutative property of addition and multiplication below.

(1) $3 + 4 = 4 + 3$ $7 = 7$

(2) $5 \times 6 = 6 \times 5$ $30 = 30$

addition and multiplication are commutative

(3) $8 - 5 = 5 - 8$ $3 \neq -3$

(4) $10 \div 5 = 5 \div 10$ $2 \neq \frac{1}{2}$

subtraction and division are *not* commutative

1. 5 + (3 + 2) = (5 + 3) + 2

2. 2 x (2 x 5) = (2 x 2) x 5

3. 8 − (4 − 2) = (8 − 4) − 2

8 − 2 = 4 − 2

4. 24 ÷ (12 ÷ 2) = (24 ÷ 12) ÷ 2

24 ÷ 6 = 2 ÷ 2

The associative property of addition and multiplication: when you add or multiply, it doesn't matter how you group the numbers (which numbers you calculate first).

1 $5 + (3 + 2) = (5 + 3) + 2$ $10 = 10$

2 $2 \times (2 \times 5) = (2 \times 2) \times 5$ $20 = 20$

addition and multiplication are associative

3 $8 - (4 - 2) = (8 - 4) - 2$ $6 \neq 2$

4 $24 \div (12 \div 2) = (24 \div 12) \div 2$ $4 \neq 1$

subtraction and division are *not* associative

Find the missing values.

1   25 + 18 = ☐ + 25   **2**   ☐ x 87 = 87 x 34

3   5 + (13 + 7) = ☐   **4**   (5 + 13) + 7 = ☐

5   17 x (2 x 5) = ☐   **6**   (17 x 2) x 5 = ☐

The zero property
Read the definitions and then fill in the blank.

Zero property of addition: the sum of any number and zero is the original number.

 5 + 0 = 

2 0 + ___ = 14 ?

Zero property of multiplication: when a number is multiplied by zero, the answer is zero.

3 9 x 0 = 

4 0 x 24 = 

"Distribute" each number to show its value.

Fill in the blanks. You are distributing the values into 100s, 10s and 1s.

38 = 30 + 8

95 = ☐ + ☐

267 = 200 ☐ + ☐

986 = ☐ ☐ + ☐

Distributive property is used to calculate
the area of this rectangle.

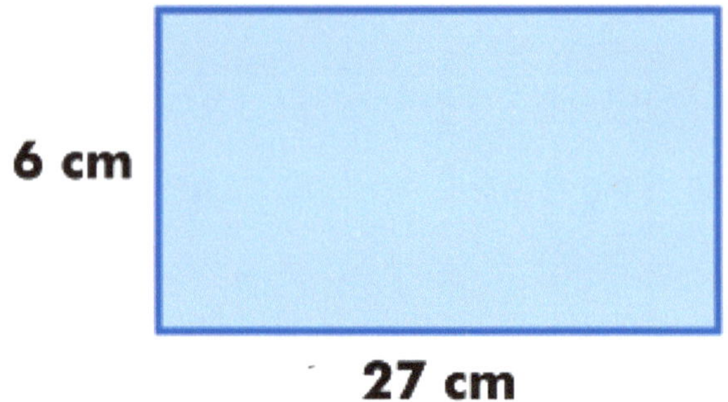

Step 1
Divide distribute 27 into tens and ones

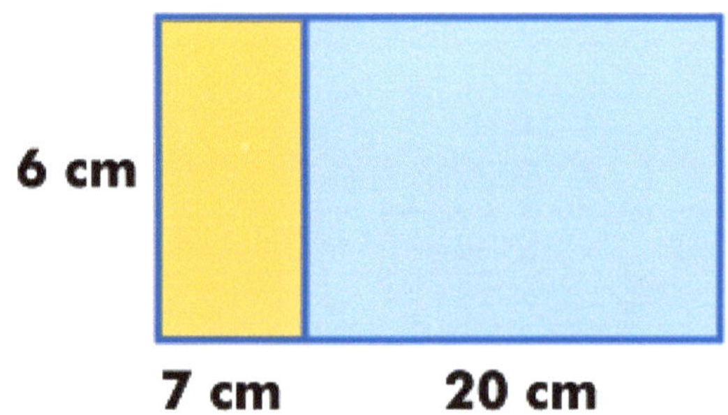

Step 2
Create the equation for the distributed values; instead of 6 x 27, we distributed the values of 27 so we now have 6 x 7 and 6 x 20. Discover solving the problem using distributive property below.

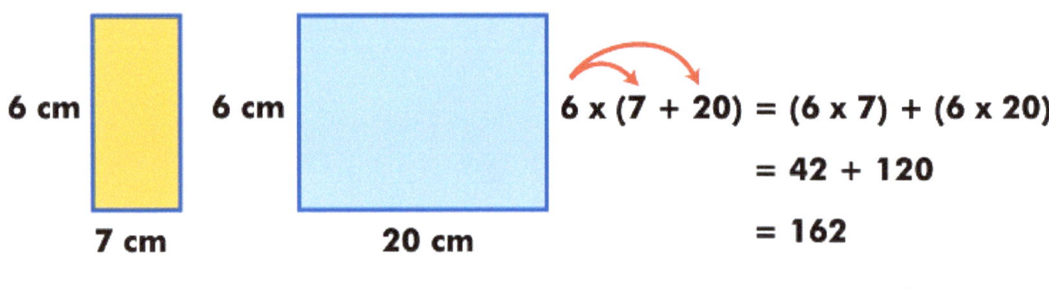

Find the missing values.

1) 8 x (6 + 2) = (8 x ___) + (8 x ___)

2) 5 x (___ − ___) = (5 x 7) − (5 x 2)

3) (11 x 4) + (11 x 3) = 11 x (___ + ___)

4) (___ x 3) + (___ x 7) = 9 x (3 + 7)

Name________________________________

Number Properties Quiz

1   **True or false? 70 + 8 can also be written as 708**

2   **Complete the equation 7 x (10 − 6) =**

    A   **(7 x 10) − (7 x 6)**

    B   **(7 x 10) − 1**

    C   **(7 x 10) + (7 x 6)**

    D   **(70 − 6)**

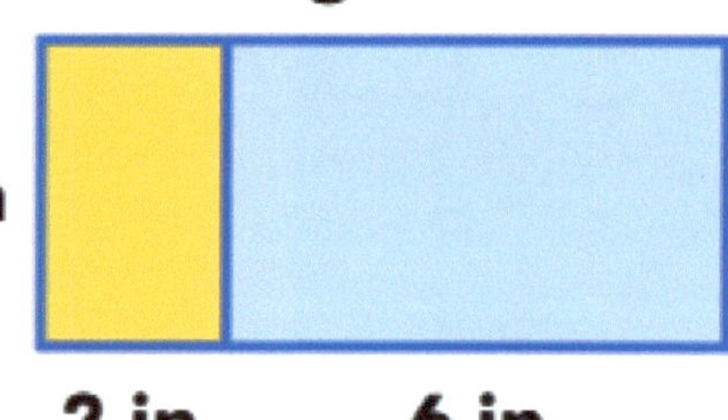

3   **(9 x 2) + (9 x 6) = ? x (2 + 6)**

4   **Find the area of Figure 1 (above) in square inches**

Rounding Numbers

Draw the numbers on the number line.

Draw these numbers on the number line. If the number is closer to 500 draw it in red, if the number is closer to 600, draw it in blue. Since 550 is halfway between 500 and 600, you'll have to use rounding rules to decide whether to choose red or blue.

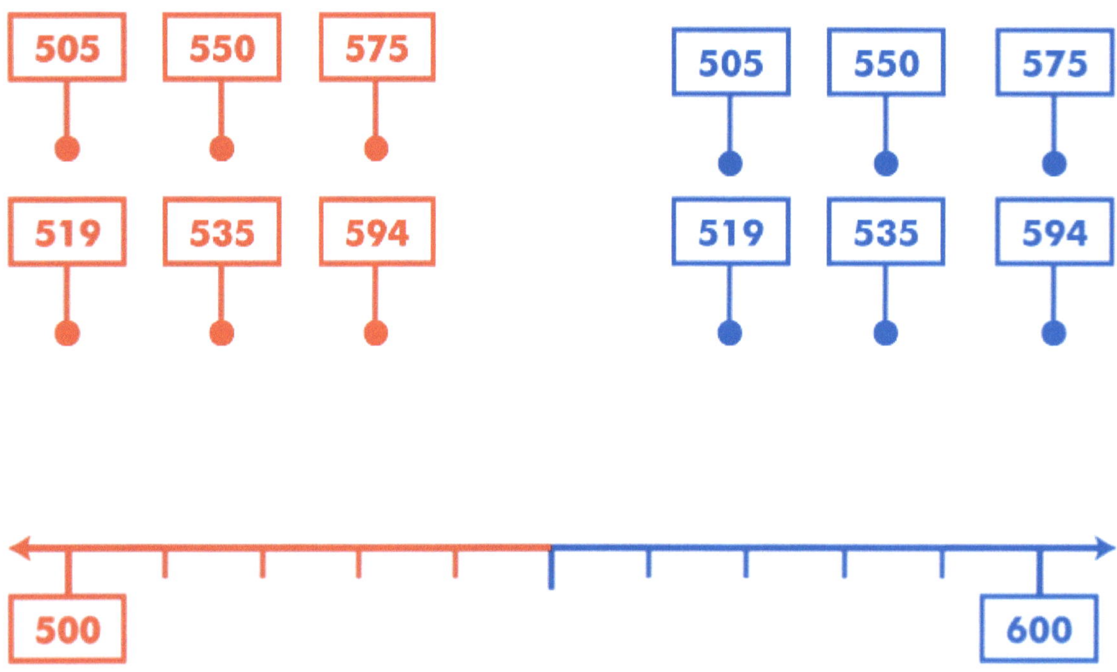

Rounding is like this beach ball on a wavy line. Which way would it naturally drop?

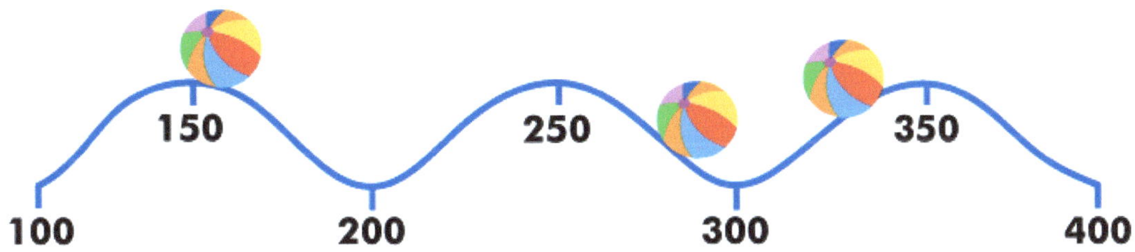

Which way would the ball naturally drop?

____

Round to the nearest 100.

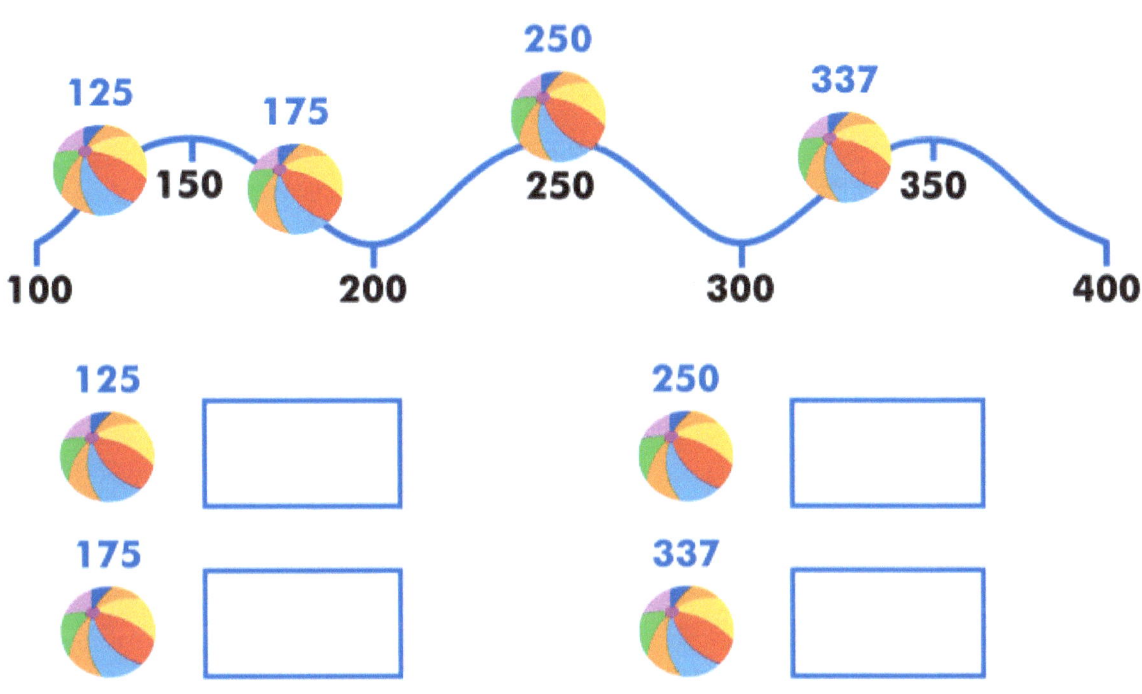

Rounding to the nearest 100

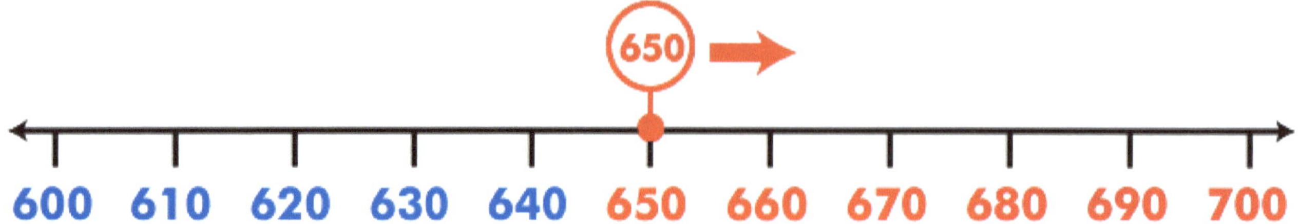

If the digit in the tens place is 0, 1, 2, 3 or 4, round down.

If the digit in the tens place is 5, 6, 7, 8 or 9, round up.

Rounding to the nearest 1,000

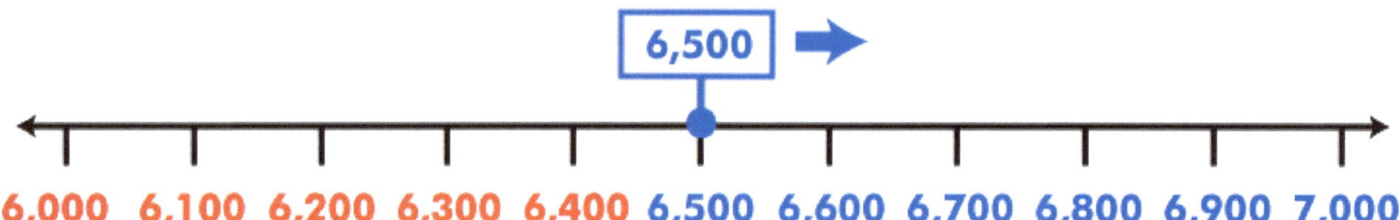

If the digit in the hundreds place is 0, 1, 2, 3 or 4, round down.

If the digit in the hundreds place is 5, 6, 7, 8 or 9, round up.

Practice what you learned.
Round to the nearest 10,000. The hockey stadium is already completed as an example.

Draw the ball on the number line rounding to the nearest 10,000.

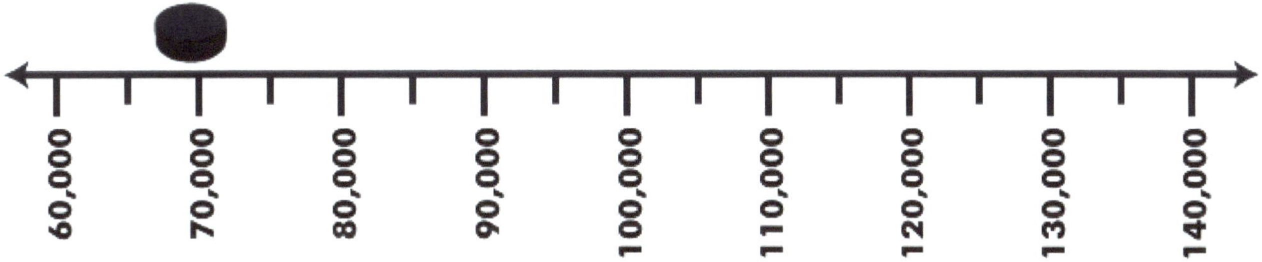

Stadium	Capacity	
Grizzly Arena	65,499	
Rooster Stadium	131,759	
Wildcat Park	75,000	
Beaver Bowl	64,599	
Bobcat Field	100,215	

Name______________________________

Rounding Numbers Quiz

1 **True or false? 4,345 rounded to the highest place value is 3,000.**

2 **Which number below would not round to 800?**

   A **789**

   B **801**

   C **849**

   D **739**

3 **Round 321 to the nearest 100.**

4 **Round 16,700 to the nearest 10,000.**

Judging Reasonable Answers

Key Vocabulary

estimate

nice number

reasonable answer

Use estimation to find "reasonable" answers.
Place check mark in the box if it is a reasonable answer and an X if it is not a reasonable answer.

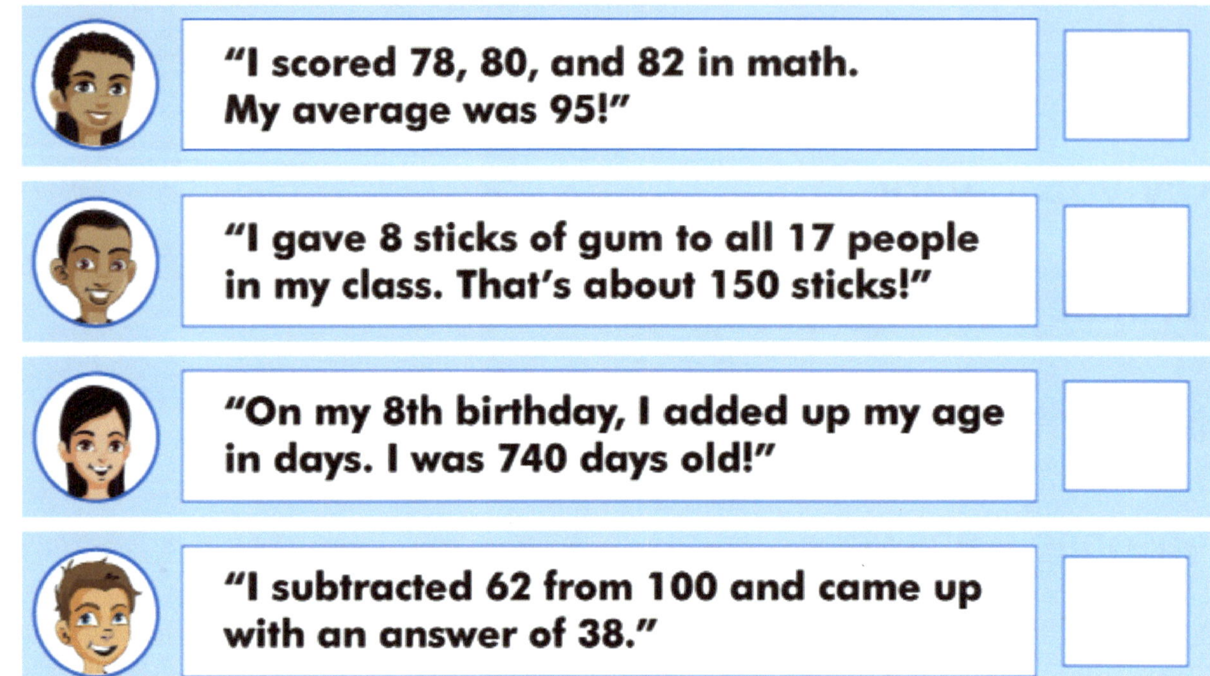

Make and estimate using nice numbers.

Round the numbers in red by drawing them the number line to the nearest 10.
This will give you "nice numbers" that are easy to add.

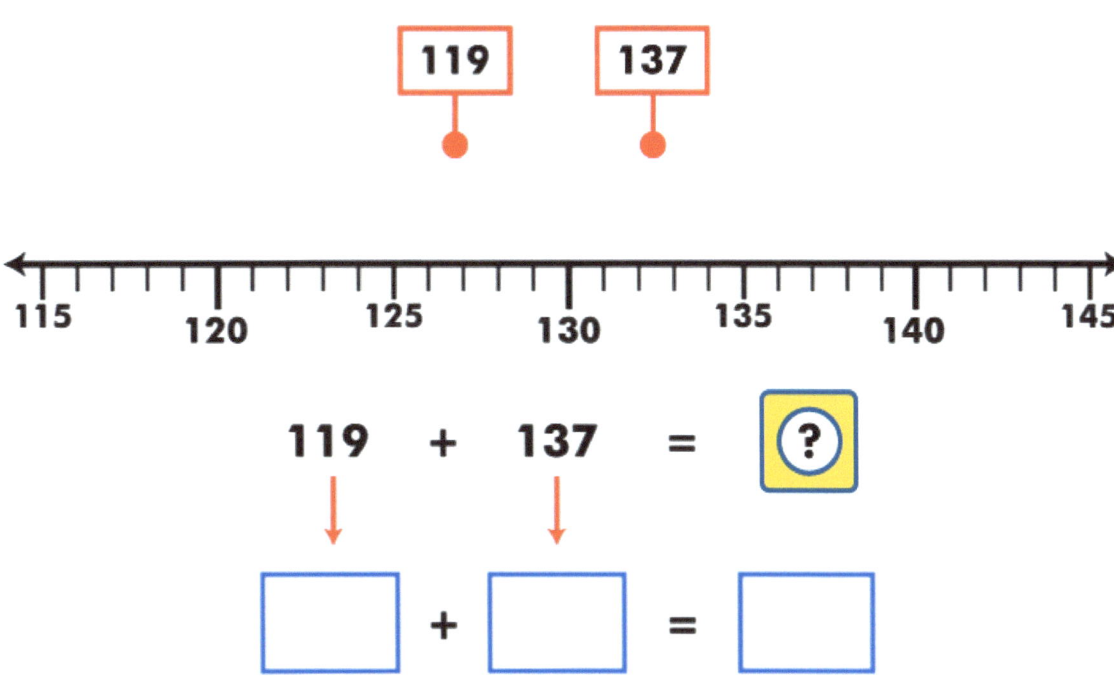

Use number friends to calculate the total points scored by Owen.
Number friends are numbers that when added together make nice numbers. The number friends are color coordinated below.

 **Points scored by Owen** 

Game	Points
1	15
2	10
3	40
4	25
5	35
6	25
TOTAL	

Use estimation to judge reasonable answers.
Fill in the blanks with R or N.

5 x 62 = 30 ☐

765 − 567 = 450 ☐

43 + 28 + 97 = 130 ☐

3 x 78 = 234 

49 − 24 + 49 = 25 ☐

812 ÷ 9 = 90 ☐

Reasonable   **N**ot reasonable

Owen, Fernando and Mia each want to spend about $6 on lunch.
Can you select lunch items for each person so that your estimate is about $6

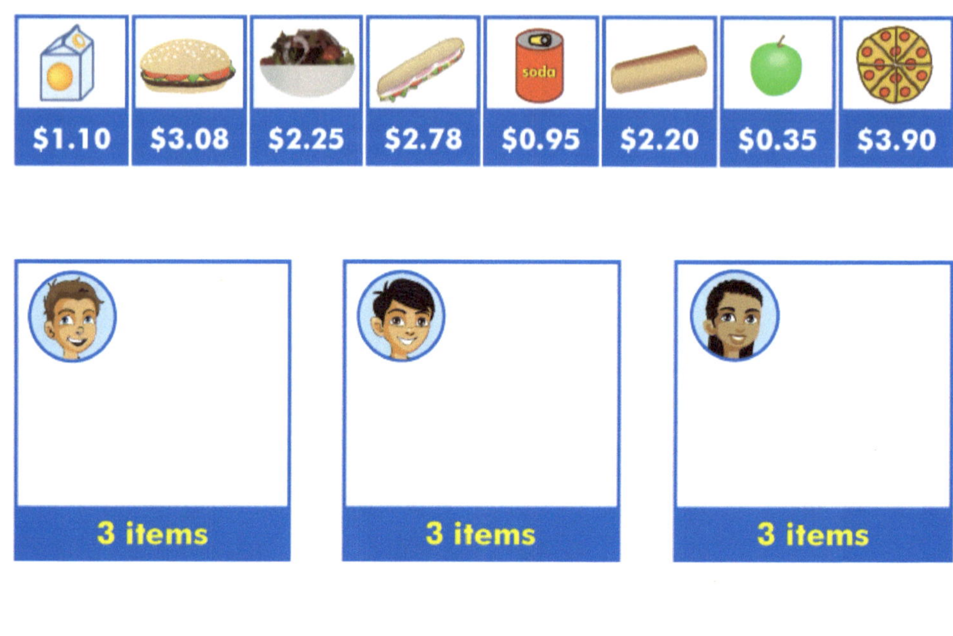

Make reasonable estimates below.

There are 345 girls on 5 buses.	
	girls per bus

Grandma drove for $5\frac{1}{2}$ hours at 37 mph.	
	total no. miles

Jack paid $3.50 for 5 bottles of water.	
	cost per bottle

It's 3,579 miles to Miami. We've driven 2,856 miles.	
	no. miles to go

Name_______________________________

Judging Reasonable Answers Quiz

1 **True or false? If you drive for 10 hours at 53 mph, you will have driven more than 500 miles.**

2 **A golfer shoots 68, 72, and 70. What is his average?**

A 68

B 69

C 70

D 71

3 **What's the best estimate for 62 x 29 = ?**
1,800   1,200   1,500   2,000

4 **What's the best estimate for 789 ÷ 8 = ?**
50   80   100   150

Newburyport, MA 01950

1-800-596-3175

OnBoard Academics employs teachers to make lessons for teachers! We create and publish a wide range of aligned lessons in math, science and ELA for use on most EdTech devices including whiteboard, tablets, computers and pdfs for printing.

All of our lessons are aligned to the common core, the Next Generation Science Standards and all state standards.

If you like our products please visit our website for information on individual lessons, teachers licenses, building licenses, district licenses and subscriptions.

Thank you for using OnBoard Academic products.